YOUR KNOWLEDGE HAS VALUE

Assigning of values of the prime number counting function to Berhard Riemann's zeros. Concept of dirichlet lines in the complex plane

William Fidler

Bibliographic information published by the German National Library:

The German National Library lists this publication in the National Bibliography; detailed bibliographic data are available on the Internet at http://dnb.dnb.de.

ISBN: 9783346583758
This book is also available as an ebook.

Print and binding: Books on Demand GmbH, Norderstedt, Germany
Printed on acid-free paper from responsible sources.

The present work has been carefully prepared. Nevertheless, authors and publishers do not incur liability for the correctness of information, notes, links and advice as well as any printing errors.

GRIN web shop: https://www.grin.com/document/1168625

Dirichlet lines in the complex plane and the assigning of values of the prime number counting function to Riemann's zeros

W M Fidler

Abstract

A procedure is developed here, by means of which, the zeros of Riemann's zeta function in the so-called Critical Strip of the complex plane, may be assigned values of the prime number counting function.

The procedure is novel and uses the concept of a Dirichlet line in the complex plane and a quantity called a nearodd (both of which are defined in the text) The process may be rendered self-contained, in the sense that, when it is associated with Gram's series the only input required is the magnitude of the imaginary part of the function $s = \frac{1}{2} + iy$, which will locate a Riemann zero on the Critical Line; vast numbers of zeros may be accessed in [2].

Further, it is shown that the Riemann conjecture is irrelevant in the assigning of any particular value of the prime number counting function to the corresponding Riemann zero.

It is suggested, pace Wiles, who obtained a proof of Fermat's Last Theorem as a by-product of his verification of the Taniyama-Shimura conjecture, that, in the light of Gödel's incompleteness theorems, Riemann's hypothesis may be undecidable.

Contents List

Introduction

Given their importance as the building blocks of all of the integers, the prime numbers, their number and disposition within the integers have been studied for millennia. In 1859 Riemann published a short paper (which only extended to six manuscript pages) concerned with an investigation of the number of prime numbers less than any given number. The outcome of the work revolutionised number theory and in the years following has resulted in almost what could be termed an industry in a particular aspect of his work, now called the Riemann Hypothesis.

Riemann showed that all the zero values of the zeta function in the positive part of the complex plane should lie in a region between $x = 0$ and $x = 1$, and, in particular, conjectured that they all should lie on the line of symmetry $x = \frac{1}{2}$.

This conjecture, which is the basis of many other conjectures in Number Theory is considered by many mathematicians to be the currently greatest unsolved problem in Mathematics—so much so that the Clay Mathematical Institute of Boston, Mass. has offered one million dollars to anyone who can produce a solution.

No attempt is made here to verify the hypothesis, for its validity is not required.

Analysis

The following has appeared in [1]. However, for a reader meeting this for the first time it is considered apposite that, for the purpose of clarity, some of the working therein should be set out again in detail.

The Riemann zeta function, $\zeta(s)$ is an extension to the series:

$$\zeta(s) = \frac{1}{1^s} + \frac{1}{2^s} + \frac{1}{3^s} + \frac{1}{4^s} + \frac{1}{5^s} + - - - - \text{---------------} \quad (1).$$

Here, the real number exponent is replaced with a complex number, $s = x + i\,y$.

It should be noted that we use Riemann's notation for the complex number but the normal mathematical notation for its real and imaginary parts.

Under the same constraint as above we write the Dirichlet eta function, $\eta(s)$ as :

$$\eta(s) = \frac{1}{1^s} - \frac{1}{2^s} + \frac{1}{3^s} - \frac{1}{4^s} + \frac{1}{5^s} - \text{------------------------} \quad (2).$$

From equations (1) and (2) we get:

$$\zeta(s) - \eta(s) = 2^{1-s}\left[\frac{1}{1^s} + \frac{1}{2^s} + \frac{1}{3^s} + \frac{1}{4^s} + - - - -\right] = 2^{1-s}\,\zeta(s).$$

For reasons which will become apparent later in the analysis we write the above as:

$$-\eta(s) = (2^{1-s} - 1)\,\zeta(s) \text{---------------------------------}(3).$$

Now, Riemann's functional equation is: $\zeta(s) = 2^s \pi^{s-1} \sin\left(\pi\,^s/_2\right) \Gamma(1 - s)\zeta(s - 1).$

It then follows that we may write the Dirichlet functional equation in terms of the Riemann functional equation

Hence, $\eta(s) = (1 - 2^{1-s})2^s \pi^{s-1} \sin\left(\pi\,^s/_2\right) \Gamma(1 - s)\zeta(s - 1).$ ---------------- (4)

We now seek a solution to equation (1) when $\zeta(s) = 0$.

Equation (1) is written out <u>in extenso</u> :

$$\zeta(s) = \frac{1}{1^{x+iy}} + \frac{1}{2^{x+iy}} + \frac{1}{3^{x+iy}} + \frac{1}{4^{x+iy}} + \frac{1}{5^{x+iy}} + \text{------------------------} \quad (5).$$

For the sake of illustration consider the second term of the above.

i.e. $\frac{1}{2^{x+iy}} = \frac{e^{-iy\,ln2}}{2^x}$, which by Euler's theorem may be written:

$\frac{1}{2^{x+iy}} = \frac{1}{2^x}[cos(y\,ln2) - i\,sin(y\,ln2)].$

Hence, we may collect terms in equation (5) and write:

$$\zeta(s) = \sum_{k=1}^{k=\infty} 1/_{k^x} \cos(y\ lnk) - i \sum_{k=1}^{k=\infty} 1/_{k^x} \sin(y\ lnk) \text{ -------------- (6).}$$

If we set $y\ lnk = k\ \pi$ it then follows that all of the imaginary terms disappear. This leaves the real part to be given by: $\sum_{k=1}^{k=\infty} 1/_{k^x} \cos(k\ \pi)$.

Hence, the real part of equation (6) becomes : $- 1/_{1^x} + 1/_{2^x} - 1/_{3^x} + 1/_{4^x} - 1/_{5^x}$ ------

This may be written: $- [1/_{1^x} - 1/_{2^x} + 1/_{3^x} - 1/_{4^x} + 1/_{5^x}$ ------].

But, this is equal to $-\eta(x)$.

If the real part of (6) is to disappear then from equation (3) we have:

$$-\eta(x) = (2^{1-x} - 1)\ \zeta(x) = 0.$$

Either the first term on the RHS disappears or the second.

The first term will vanish if **x = 1**, but this would reduce equation (1) to the Harmonic series which is known to diverge, albeit slowly, to infinity. The product above would then be of the form: **0** . ∞, which is indeterminate. It then follows that we must take $\zeta(x)$ to be zero.

We now return to the functional forms for ζ and η.

Again, Riemann's functional equation is: $\zeta(s) = 2^s \pi^{s-1} \sin(\pi\ ^s/_2)\ \Gamma(1 - s)\zeta(s - 1)$.

Now, the sine term will vanish if we set **s = -2n**, where **n** is a real integer.

It then follows, that setting **s = -2n** will yield the following result: $\zeta(-2n) = 0$, and hence, from equation (3), $\eta(-2n)$ will also vanish. It is important to emphasize that the sine term will vanish if, and only if, n is an integer.

This procedure outlined is, of course, that employed in generating the so-called 'trivial zeros' of the Riemann zeta function. Further, from the functional equation we see that whatever ζ is evaluated to on the left of **s** = ½ is determined by its evaluation on the same point reflected across **s** = ½. This, together with the analytic continuation of η provides the ability to compute ζ anywhere in the complex plane. In addition, since $\zeta(s)$ has no zeros to the right of **Re(s) = 1**, then the functional equation predicts that there are no other non-trivial zeros to the left of **Re(s) = 0** and hence all of the non-trivial zeros lie within the Critical Strip; this will be used later in the analysis.

Dirichlet lines in the complex plane

It follows from the preceding that, in the complex plane, along lines with an ordinate given by the formula: $y = k\pi/lnk$, for any given constant k, where, of course k is an integer, the imaginary part of the zeta function is zero at all points. Along such a line the real part of the zeta function at any point is shown to be equal to the negative of the Dirichlet function for a real number.

We name these lines Dirichlet lines in honour of Dirichlet, for they bear, in effect, the distribution of the Dirichlet function $\eta(x)$ in the range, $-\infty \leq \eta(x) \leq +\infty$. Further, we see that the Riemann zeta function is zero whenever a Dirichlet line intersects a 'critical line' through a trivial zero of the zeta function. These remarks apply to all k.

The assigning of values of the prime number counting function to the Riemann zeros along the line passing through s = ½

From [2] we obtain the magnitude of the imaginary part of the coordinate of the first Riemann zero and which we truncate to **14.134725**, although the decimal part of the actual number is calculated to one thousand places. Whilst this level of accuracy is deemed necessary to ensure that the imaginary part of the zeta function is zero, or very close thereto, such precision is not required to associate a value of the prime number counting function with a Riemann zero. In the following illustration it is shown how a value of the prime number counting function may be linked to a zero of Riemann's zeta function

Let it be assumed that the magnitude of the imaginary part of **s** may be calculated from the simple formula: $y = k\pi/lnk$, where y = 14.134725 = Im(s).

We rearrange the above equation in iterative form, i.e. $k_{n+1} = (y/\pi)lnk_n$ ---------------- (7).

Hence for the example here, $k_{n+1} = 14.134725/\pi . lnk_n = \Omega \, lnk_n$ --------------------(8).

Using an initial value of 10, equation (8) iterates rapidly to yield k = **10.637895**, although, all of the digits in the decimal part are superfluous for our purpose. It is found that all subsequent iterative calculations shown herein may be started with a value of, $k_1 = 10$..

It follows that the point in question is 'bracketed' by the Dirichlet lines for k = 10 and k = 11 which intersect the so-called Critical Line at s = x = ½ at $R(s) = -\eta(1/2)$ as shown in Fig1.

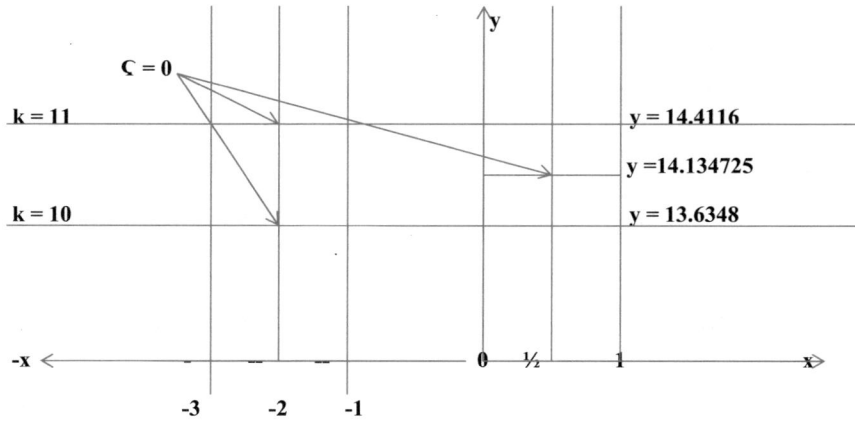

Fig1

Now, the integers, **k = 10** and **k = 11** each have an associated value of the prime number counting function (indeed, they may be the same). Further, we discard the upper value of **k** for it makes no sense to evaluate the number of primes associated with the point of interest because the location of the Riemann zero there is less than that for the zeta function at **k = 11**. In addition, the lower bound is an even number and there are, with the exception of **2** no even prime numbers. Hence, for the purpose of associating a value of the prime number counting function we take the lower nearest odd number (which may, or may not, be prime), and which, in this case is **9** ($y = 12.8682$). For brevity we will, in all further illustrations refer to a number of this nature as a **nearodd** and denoted by the symbol, **d**. Thus, every 'Riemann zero' and its associated value of the prime number counting function will have a corresponding nearodd which is located within the matrix, Table1, developed in [3] (where the primes are presented in bold and underlined), or, from Kulsha's data [4].

We then find, from Kulsha's data that the prime number counting function value associated with the nearodd of the first Riemann zero is equal to **4**, and this is verified by inspection of the matrix, which only contains the odd numbers and so does not take account of the first prime, **2**. Note that the lowest row of numbers denotes the column numbers of the matrix.

3	5	7	9	11	13	15	17	19	21	23	25	27	29	31
33	35	37	39	41	43	45	47	49	51	53	55	57	59	61
63	65	67	69	71	73	75	77	79	81	83	85	87	89	91
93	95	97	99	101	103	105	107	109	111	113	115	117	119	121
123	125	127	129	131	133	135	137	139	141	143	145	147	149	151
153	155	157	159	161	163	165	167	169	171	173	175	177	179	181
183	185	187	189	191	193	195	197	199	201	203	205	207	209	211
213	215	217	219	221	223	225	227	229	231	233	235	237	239	241
243	245	247	249	251	253	255	257	259	261	263	265	267	269	271
273	275	277	279	281	283	285	287	289	291	293	295	297	299	301
1	2	3	4	5	6	7	8	9	10	11	12	13	14	15

Table 1

Now, it was noted in [5] that Riemann's prime number counting function **R(x)** was an exceptionally good approximation to the prime number counting function, $\pi(x)$. It was proved by G H Hardy in 1914 that **R(x)** was equivalent to Gram's series, **G(x)**. Hence, using the concept of the nearodd we may write:

$$R(d) = G(d) = 1 + \sum_{k=1}^{\infty} \frac{(\ln d)^k}{k\, k!\ \zeta(k+1)}.$$

This function converges rapidly, but requires extensive computing power, for the product **kk!** rapidly becomes very large even for moderate k, and, in addition, the zeta function must be calculated with great precision.

Nevertheless, in principle, we can, through the nearodd, use this function to associate any Riemann zero with a value of the prime number counting function. In order to engage with the matrix, we omit the first term on the RHS of the above equation.

In Table2 we show the first ten Riemann zeros from [2] with their corresponding values of nearodd, **d** calculated from equation (7), and the magnitudes of the corresponding prime number counting function from [4]. Further, to demonstrate the accuracy of the above formula (which is quite astonishing), Table3, from [6] and unattributed there, is presented below.

Im(s)	Ω	d	R(d)
14.1347	4.4992	9	4
21.022	6.69151	19	8
25.018	7.9612	25	9
30.42487	9.68453	33	11
32.93506	10.48355	37	12
37.5861	11.96402	45	14
40.91871	13.02483	51	15
43.32707	13.79143	55	16
48.00515	15.28051	63	18
49.77383	15.8435	65	18

Table2

x	$\pi(x)$	R(x)
100,000,000	5,761,455	5,761,552
200,000,000	11,078,937	11,079,090
300,000,000	16,252,325	16,252,355
400,000,000	21,336,326	21,336,185
500,000,000	26,355,867	26,355,517
1,000,000,000	50,847,534	50,847,455

Table3

It should be noted that if we are using Kulsha's data then, $R(d) = \pi(d)$.

The values in the fourth column of Table2 are the number of primes in the range of the respective nearodd, **d**.

That this is the case may be verified from inspection of that part of the matrix of odd numbers shown in Table1, remembering that the actual number of primes calculated is one greater than the number of primes counted there, for the matrix contains only odd numbers.

The determination of a nearodd is easy, for equation (8) converges very rapidly.

The reader is invited to determine and verify the magnitude of **R(d)** in conjunction with equation (8) and Kulsha's data plus the data presented in Table1, or extensions thereof.

To emphasise the veracity of the method of iteration we choose, at random, a rounded value of the 29th Riemann zero from [2], i.e. **98.8311942.** Hence, $\Omega = $ **98.8311942**$/\pi = $ **31.458946**.

With this value as parameter equation (8) iterates to: **k = 159.57614,** and so **d = 159.**

From Kulsha's data **R(d) = R(159) = 37;** counting the primes in the matrix gives **R(159) = 36,** but again, the matrix contains only the odd numbers and so omits the first prime,**2.**

A graph for the variation of the prime number counting functions of the first 25 Riemann zeros with the corresponding nearodd is shown in Fig2

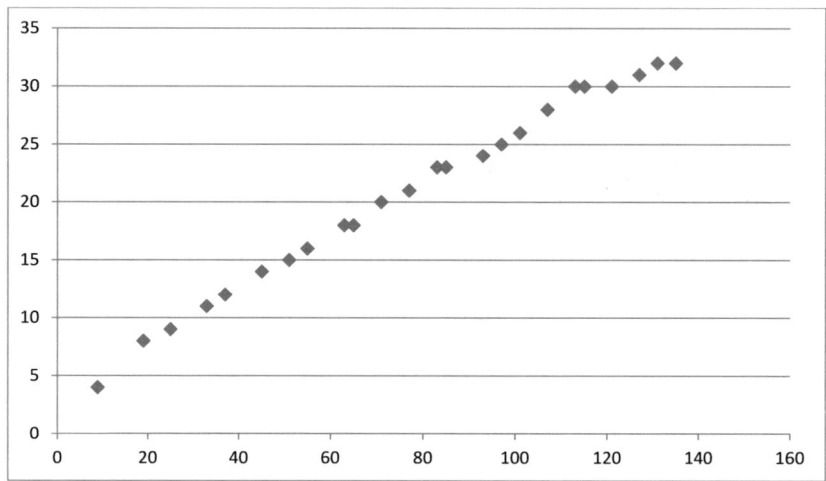

Fig2

Since we have used Kulsha's data in the determination of the above, then this graph represents, without approximation, the variation of the prime number counting function in the range shown.

We now present in summary form the procedures for assigning values of the prime number counting function to Riemann's zeros, in particular those on **s = 1/2,** for the zeros on the lines of symmetry passing through the trivial zeros do not require iteration to establish their location in the range of the counting numbers.

The parameter, **Ω,** is formed by dividing **Im (s)** (suitably truncated) from [2], by π

.Equation (7) is iterated to convergence. This will give a result of the form **w** + **f**, where **w** and **f** are the whole and decimal parts of the number, respectively. It should be noted that **f** can never be zero for this would mean that the zero was located at the intersection of a Dirichlet line and the Critical Line, and here, whilst **ImR(s)** = **0,** the real part of the zeta function is $-\eta(\frac{1}{2})$, where η is Dirichlet's alternating function.

If **w** is odd then the nearodd, **d** = **w.** If **w** is even then **d** = **w** – **1.**

The value of the prime number counting function **R(d)** is then found from:

$$R(d) = G(d) = \sum_{k=1}^{\infty} \frac{(\ln d)^k}{k\,k!\,\zeta(k+1)} \quad \cdot$$

The leading term of Gram's series is omitted in this case so that we may investigate the distribution of the prime number counting function in the matrix of odd numbers, which, of course, must contain all of the prime numbers, with the exception of **2.**

The Riemann Hypothesis

Riemann's hypothesis states that all of the non-trivial zeros of the zeta function lie in the Critical Strip and on the line s = ½. In 1921 it was shown by Hardy and Littlewood [6] that there are an infinite number of zeros on the so-called Critical Line (s = ½). We have shown [1] that there are an infinite number of zeros in the negative half of the complex plane on the vertical lines passing through the trivial zeros at $x = -2n$, for n = 1, 2, 3, etc. We have no intention of engaging in a Cantorian bidding war to establish which of the above regions of the complex plane has the most number of zeros.

In view of the mathematical and numerical firepower that has been levelled to date against Riemann's conjecture, to no avail, it is considered by the author that, whilst probably true, the veracity thereof, may be, in the light of Gödel's incompleteness theorems, undecidable, for these are two theorems of mathematical logic that demonstrate the inherent limitations of every formal axiomatic system capable of modelling basic arithmetic. However, in the light of the comparisons between $\pi(x)$ and R(x) shown in Table3, we take Riemann's prime number counting function and it's equivalence to Gram's series to yield the correct number of primes in any given range.

Now, we showed in [1] that there are an infinite number of Riemann's zeros in the 'critical strips' on the lines passing through the trivial zeros, and here we see that each zero is located at the intersection of a Dirichlet line with a critical line. The determination of the location, y (height) of a zero in these strips is given by the simple formula: $y = (k\pi)/\ln k$. In the right hand half of the complex plane we used the same formula to determine the height of the first 'non-trivial' Riemann zero, but , in this instance the above formula was rearranged into the iterative form shown in equation (8).

Consider Fig3.

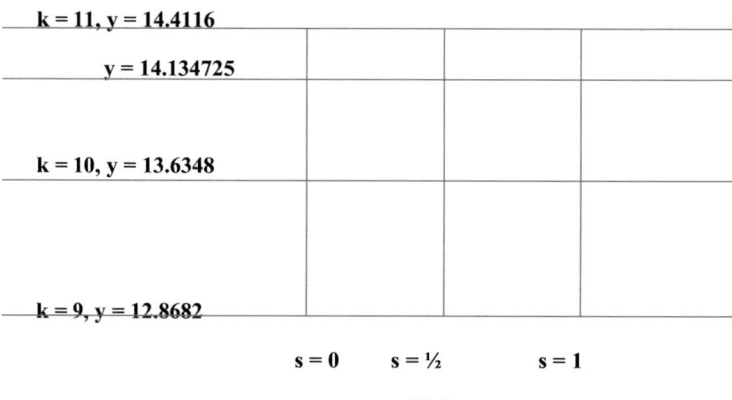

k = 11, y = 14.4116

y = 14.134725

k = 10, y = 13.6348

k = 9, y = 12.8682

s = 0　　　s = ½　　　s = 1

Fig3

We showed that the nearodd corresponding to $y = 14.134725$ is $k = 9$. It then follows that we have, in this instance, determined the nearodd for **all of the Riemann zetas, including all of any zeros that lie in the region $0 \leq s \leq 1$ and $12.8682 \leq y < 14.4116$.** That there is a singularity at $x = 1$ is of no consequence.

In the use of the Gram series we now see, given that the formula may be computed to convergence, that we must use a 'proper' number as the argument, and that this number is a nearodd. The utility of the imaginary part **Im(s)** of the location of the Riemann zeros now becomes apparent for, using equation (7), with $\Omega = \dfrac{Im(s)}{\pi}$ we can determine all of the nearodds for not only the Riemann zeros, but also for any other point in regions such as that exemplified above.

In [1} it was argued that each 'critical strip' in the negative half of the complex plane had its own prime number counting function. This was realized by an ad hoc scaling of the integer, **k** in the expression for the Gauss/Legendre prime number counting function. The scaling factors were the reflections of the trivial zeros in the line of symmetry, $s = \frac{1}{2}$, and so, introduced the zeta function into the prime number counting function. In the light of the work presented here it is seen that whilst the latter may be of some mathematical interest, in that the scaling factor ranges from Apery's constant, to unity at infinity and would introduce changes in the prime number counting function by the introduction of the scaled counting function at appropriate places in an attempt to bring the number of primes closer to $\pi(x)$, at bottom it had no legitimate mathematical basis for it resulted in a different prime number counting function in each 'critical strip'.

In contradistinction, the method developed here shows that the prime number counting function is, as it should be, the same throughout the complex plane.

The Critical Strip

Since there are an infinite number of Dirichlet lines in the complex plane then the Critical Strip may be divided into an infinite number of cells of the form described previously. All of the cells are bounded by Dirichlet lines which have odd numbers as parameters. The magnitude of **Im(s)** will determine the cell in which it lies. Indeed, there are an infinite number of Riemann zeta functions in any given cell, none, or some of which, may be a zero; this is immaterial for all of them have the same nearodd.

The linking of the Riemann zeros by the method developed here to values of the prime number counting function is completely independent of where these zeros are located in the Critical Strip. It then follows, that here, Riemann's conjecture is irrelevant

In order to correspond to the numbers in the part of the matrix shown in Table1 we choose the first cell to be bounded by the numbers **3** and **5**.

It should be emphasized that, although the difference between the parameters of the Dirichlet lines which form the upper and lower boundaries is the same for all cells, this should not be taken to imply that the difference between the vertical heights of the cells are the same.

Discussion

The utility of the Dirichlet (D) lines cannot be overstated for there is such a line for each natural number, and that number is a parameter of the line.

It is seen that the magnitude of the zeta function is zero at the intersection of a D- line with the vertical lines through the trivial zeros of Riemann's zeta function. In all other intersections of a D-line with a vertical line the magnitude of the Riemann zeta function is real and is the negative of the Dirichlet eta function of the abscissa of that intersection, i.e. $R(s) = -\eta(x)$.

This has a rather interesting consequence. Since there is a D-line for each natural number it then follows that there is a D-line at infinity for we have shown [1] that this is the D-line with parameter, 1 which will intersect the abscissa, x = ½. But, the negative of the Dirichlet eta function for x = ½ is **-0.6049** (to four decimal places), and so this precludes the presence here of a Riemann zero at infinity. Of course, this implies nothing regarding the number of Riemann zeros on the line x = ½, although the number of zeros on the vertical lines passing through each of the trivial zeros [1] is infinite, for each zero there is associated with its own D-line.

In contradistinction to the objective of the work in [7], which was concerned with the disposition of the prime numbers and used Kulsha's data, here, the process may be rendered self-contained if we are willing to use Gram's series to determine the magnitude of the prime number counting function for any **d** , for we may then dispense with Kulsha's data, and all that is required to associate a Riemann zero with the value of a particular prime number counting function is the magnitude of that Riemann zero, the number of which, it has been reported, extend to 10^{36}. Indeed, if Gram's series is regarded as a formula which will return a result independent of the nature of the argument (i.e. a whole number or otherwise) then, we may argue from the perspective of Table3 that the use of a nearodd as argument will yield a smaller result than the use of the number (k_{n+1}) itself and so may move the number of primes closer to that given by $\pi(d)$.

In view of the arguments presented here, the value of the imaginary part of the location of a Riemann zero lies in the yielding of the associated nearodd and hence the corresponding prime number counting function. This observation is validated by the 'directional' nature of the Euler product formula which states that the zeta function is related to the product of the primes.

The cells themselves require some comment. Unity is not a prime number, and in any case, as noted above, the D-line with unity as parameter passes through infinity. The origin of the cells cannot start at **2** for this is unique in the sense that it is the only prime number which is an even number and so cannot be designated a nearodd. Hence the first cell has the D-line with parameter **3** as its lower boundary. As noted previously, all of the infinity of zeta functions at the permitted locations in any cell may be associated with the same nearodd.

References

[1] On the Riemann Hypothesis

     W M Fidler

  GRIN Verlag, ISBN 9783346388575   (2021).

[2] Tables of the zeros of Riemann's zeta function

     A Odlyzko

  www.dtc.umn.edu/~odlyzko/zeta-tables/index.html

[3] Determining the primality of a number by the use of an accelerated version of trial division

     W M Fidler

  GRIN Verlag, ISBN 9783346493002 (2021)

[4] The fluctuations of the prime number counting function $\pi(x)$.

     Complied by A V Kulsha

  www.primefan.ru/stuff/primes/table.html

[5] The first 50 million prime numbers

     Don Zagier

     Inaugural Lecture, Bonn University, May 1975.

[6] The zeros of Riemann's zeta function on the Critical Line.

     G H Hardy & J E Littlewood

     Math.Z, 10(3 – 4): 283 – 317 (1921)

[7] On the disposition of the prime numbers.

     W M Fidler

  GRIN Verlag, ISBN 9783346548696 (2021).

W M Fidler

January 2022.